CORVETTE

Mike Mueller

MBI

This edition first published in 2004 by MBI, an imprint of MBI Publishing Company, Galtier Plaza, Suite 200, 380 Jackson Street, St. Paul, MN 55101-3885 USA

MBI titles are also available at discounts in bulk quantity for industrial or sales-promotional use. For details write to Special Sales Manager at Motorbooks International Wholesalers & Distributors, Galtier Plaza, Suite 200, 380 Jackson Street, St. Paul, MN 55101-3885 USA.

On the front cover: This 1961 fuel-injected Corvette has the optional racing equipment. The addition of the competition-conscious oversized fuel tank made the installation of the optional removable hardtop mandatory because there was no room behind the seat for the conventional convertible top. This rare Corvette is owned by Elmer Lash of Champaign, Illinois.

On the frontispiece: The most distinctive feature of 1953 through 1955 Corvettes was the beautifully detailed fence mask headlight stone guard.

On the title page: The 1968 427 Sting Ray was quickly likened to a Coke bottle with its bulging front and rear quarters and a slimmed-down midsection. This silver one belongs to Guy Landis of Kutztown, Pennsylvania. The blue 1971 454 Stingray belongs to Tom Biltcliff of Kutztown, Pennsylvania.

On the back cover: The fiberglass faithful eagerly await the arrival of the new sixth-generation Corvette. The 2005 Corvette was first unveiled to the public in January 2004.

About the author: Author Mike Mueller is a Georgia-based automotive writer and photographer, and a former staff editor for many magazines, including *Automobile Quarterly*. He has also had articles and photos published in *Autoweek, Life, Collectible Automobile, Truck Trends, Truckin'*, and *Car Craft* magazines, as well as countless muscle car calendars. He has written many titles for MBI Publishing Company including *Mustang 1964 1/2–1973*, *Corvette C5*, and *Chevrolet Pickups*.

ISBN 0-7603-1968-5

Edited by Amy Glaser
Designed by Design53, St. Paul, Minnesota

Printed in China

When the National Corvette Museum in Bowling Green, Kentucky, finally opened on Labor Day weekend in 1994, the legendary piece of Corvette history that greeted visitors at the main lobby was Zora Arkus-Duntov's ill-fated SS racer from 1957. Shown here in front of the museum, the magnesium-bodied SS was later returned to its permanent home at the Indianapolis Motor Speedway Hall of Fame museum.

Acknowledgments

It was an event seven years or four decades in the making, depending on your perspective. Many of you out there who've lived with and loved America's sports car from its humble birth in 1953 may have wondered if the long-deserved, long-awaited National Corvette Museum would ever open its doors. Originally discussed in 1987, the idea for a four-walled tribute to one of the greatest cars this country has ever produced went through more than its fair share of ups and downs before it became reality. On September 2, 1994, the ribbon across the doors of Valhalla in Bowling Green, Kentucky, was finally cut.

Everyone was there—Zora Arkus-Duntov, Dave McLellan, Dave Hill, Jim Perkins, Larry Shinoda, the Beach Boys, and about 120,000 others. Nearly all of the memorable cars were there, including Duntov's 1957 SS, Mitchell's Sting Ray racer Manta Ray, Mako Shark, Purple People Eater, Big Doggie, and the one millionth Corvette. About the only major omissions from the lineup during that first over-crowded weekend were the SR-2 and Grand Sport. But if you wanted to see the greatest collection of Corvettes ever, the National Corvette Museum's grand opening represented a once-in-a-lifetime chance to enjoy as many of them as you could possibly imagine, all together under one very high-pitched roof.

But if you were stuck at home, perhaps you could use a little sampling of what that moment was like. While this humble publication can by no means match the grandeur of the National Corvette Museum, it can offer as much rolling legend as can fit in 96 pages. Sure, not all the greats are here, but many of them are.

Special thanks go to all the folks who allowed me to feature their fine Corvettes in this book. In general order of appearance, they are:

1967 L88 and 1969 ZL1 Corvettes, Roger and Dave Judski, Roger's Corvette Center, Maitland Florida; 1968 427 Sting Ray, Guy Landis, Kutztown, Pennsylvania; 1971 454 Stingray convertible, Tom Biltcliff, Kutztown, Pennsylvania; 1990 ZR-1, Ed and Diann Kuziel, Tampa, Florida; 1955 Corvette, Elmer and Dean Puckett, Elgin, Illinois; 1956 Betty Skelton racer, 1956 SR-2, 1963 Grand Sport, 1967 L-88, 1975 convertible, and 1978 Indy Pace Car, Bill and Betty Tower, Plant City, Florida; 1957 Airbox Corvette, Milton Robson, Gainesville, Georgia; 1961 Big Tank Corvette, Elmer and Sharon Lash, Champaign, Illinois; 1963 Z06 Sting Ray (silver), Bob Lojewski, Cook County, Illinois; 1963 Z06 Sting Ray (gold), Ron Landis, Wyomissing, Pensylvania; 1965 fuel-injected Sting Ray, Gary and Carol Licko, Miami, Florida; 1965 396 Sting Ray, Lukason and Son Collection, Florida; 1967 L71 Sting Ray convertible, Chet and Deb Miltenberger, Winter Park, Florida; 1968 L89 Sting Ray, Elmer and Dean Puckett, Elgin, Illinois; 1974 454 Corvette, Robert Boynton, Jr., Palm Harbor, Florida; 1972 LT-1 Stingray, Steve and Nora Gussack, Winter Springs, Florida; 1984 serial number 00001 Corvette, Dick Gonyer, Bowling Green, Ohio; 1982 Collector's Edition, Dan Holton, Gainesville, Florida; 1996 Collectors Edition, Jim Morris, Winter Haven, Florida; 1988 35th Anniversary Corvette, Don and Denise Sanzera, Marco Island, Florida; 1991 Callaway Twin Turbo Speedster, Milton Robson, Gainesville, Georgia; 1993 40th anniversary ZR-1 Black Widow, Jerry Crews, Longwood, Florida; 1997 C5 coupe, Tom and Kelly Sellers, Champaign, Illinois.

A hearty thank you to everyone.

A lot has changed during the Corvette's half-century on the road. Chevrolet marked the car's 50th birthday with a special anniversary model in 2003.

Introduction
A Half-Century of Life in the Fast Lane

When the first Corvette rolled off its make-shift assembly line in Flint, Michigan, on June 30, 1953, it certainly fit the classic sports car mold, with its somewhat crude folding top and plastic side curtains in place of roll-up windows. All 300 first edition roadsters were painted Polo White with red interiors, and all were powered by a muscled-up version of Chevrolet's yeoman stovebolt six-cylinder, backed by a less-than-desirable two-speed Powerglide automatic.

"Limited" was a fair description in more ways than one and helped explain why the Corvette didn't take off as fast as hoped. Adding exterior paint choices in 1954 did little to change that fact. With nearly a third of the 3,640 1954 Corvettes built sitting unsold at year's end, many at Chevrolet wondered if it wasn't best to cut and run.

Thankfully they didn't, and the Corvette quickly found its niche after V-8 power was added to the mix in 1955. Nearly four decades later, Chevrolet rolled out its one millionth Corvette—an Arctic White convertible with red interior—on July 2, 1992, and proved that someone at GM knew what they were doing. Among others, names like Harley Earl, Ed Cole, Zora Arkus-Duntov, Bill Mitchell, Dave McLellan, and Dave Hill quickly come to mind.

It was longtime GM styling mogul Earl who first campaigned for the little two-seat sportster in the early 1950s. Cole supplied support for the project early on, first as Chevrolet's chief engineer, then as the division's general manager beginning in July 1956. Duntov, the so-called "father of the Corvette," needs no introduction. After Duntov retired as Corvette chief engineer in 1975, Dave McLellan took his place, followed by Dave Hill in 1992. Mitchell replaced Earl as head GM stylist in 1958 and was responsible for the startling Sting Ray transformation unveiled for 1963.

As for the four-wheeled legend that these and other great men and women helped form and fashion into what is still one of the world's greatest sporting machines, where do you begin? Perhaps discounting, with all gentleness, the comparatively weak-kneed models built in the performance-starved late 1970s and early 1980s, hasn't damn near every Corvette built been a truly great car? And hasn't Chevrolet continually outdone itself just about every year with the best 'Vette yet? Perhaps detailing milestones isn't the proper approach. Maybe a basic outline is the better choice.

For starters, the Corvette was built in St. Louis, Missouri, from December 1953 through July 1981, and moved to Bowling Green, Kentucky, in June 1981, where it is still built today. The Corvette has evolved through six definable generations, beginning with the solid-axle variants of 1953 through 1962. The so-called mid-year models of 1963-67 make up the second generation, followed by the third running from 1968 to 1982. After the hiccup year of 1983, an all-new Corvette appeared for 1984 to kick off the fourth generation, which ended in

Fuel injection appeared as a Corvette option, at a cost of $484.20, for the first time in 1957. Various modifications raised fuelie output over the years before the Rochester injection equipment was deleted after 1965.

available in 1957, the same year the enlarged 283 ci small-block V-8 and four-speed manual transmission debuted. Standard displacement grew to 327 cubic inches in 1962, when only a single four-barrel was offered atop the three available carbureted V-8s. The last fuelie Sting Ray was built in 1965 when the Corvette's first big-block V-8 appeared as the top power option. Originally offered in 396-ci form, the legendary Mk IV big-block bully was bumped up to 427 cubic inches in 1966, and then to a whopping 454 cubic inches in 1970.

Meanwhile, small-block performance was progressing at full speed as the 370 horsepower LT-1 350 was introduced for the 1970 Corvette. The most powerful carbureted small-block (the fuel-injected 327 rated at 375 horses in top tune and the ZR-1's injected LT5 later hit 405 horsepower) ever offered under that long fiberglass hood was the first-generation LT-1. It lasted three years before it fell victim to rising insurance costs, ever-tightening

1996. The C5 dominated Mainstreet U.S.A. from 1997 to 2004, and the new C6 is waiting in the wings to be released in 2004.

All Corvettes were convertibles through 1962, and an optional removable hardtop was available after 1956. Roll-up windows were added that year as well, which made Chevrolet's sports car more socially acceptable from a Yankee perspective. Independent rear suspension and a coupe model first appeared in 1963, when all Corvettes became Sting Rays—a name that stuck until 1976. From 1969 to 1976, "Stingray" was used. Beginning in 1976, all Corvettes were coupes when the droptop model was deleted, but it reappeared in 1986. Single headlamps were used from 1953 through 1957, followed by dual units in 1958. Hideaway headlights were the norm from 1963 to 2004.

As for power, Chevrolet's groundbreaking 265-ci overhead-valve V-8 replaced the valiant, yet disappointing, six in 1955. Optional dual four-barrel carbs appeared the following year. Fuel injection was

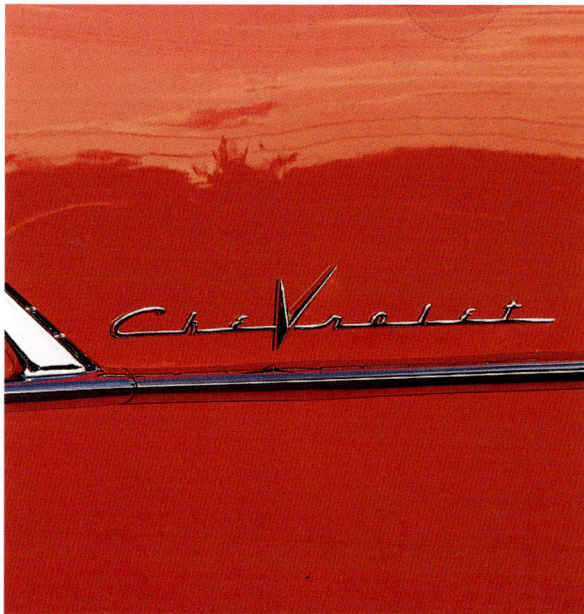

The large, gold "V" tacked over the Chevrolet emblem on this 1955 Corvette signifies the presence of a 265-ci V-8, the engine that basically saved America's only sports car from quick extinction.

"If you take that off," claimed GM styling chief Bill Mitchell, "you might as well forget the whole thing." Mitchell was referring to the stinger, his pet Sting Ray styling element that separated the all-new 1963 Corvette's rear windows. Zora Arkus-Duntov and many others didn't like the split-window theme because of the way it hindered rearward vision. Despite Mitchell's emotional appeal, Chevrolet took this stinger off for 1964 and didn't forget the whole thing.

A new Corvette body for 1968 (right) drew both raves and complaints for its sexy shape. Some said it was "painfully too American." The two big-block V-8s beneath these Corvettes' hoods are truly American in this case. The silver 1968 has the 427 derivative of Chevy's Mk IV big-block and the blue 1971 convertible has the larger 454, introduced for 1970. Big-block Corvette production lasted until 1974, when the last 454 Sting Ray was built.

Both the aluminum-head L88 (back) and all-aluminum ZL1 427s were jokingly rated at only 430 horsepower. Actual output zoomed past 500 for these race-ready rockets. Only 20 1967 L88 coupes were built, and two 1969 ZL1s escaped Chevrolet Engineering.

emissions standards, and tougher federal safety specifications. The dreaded fuel crunch of the 1970s also helped bring down the 454 Stingray when the last big-block Corvette rolled out of St. Louis in 1974.

Carburetors disappeared after 1981 as Cross Fire injection was made standard for the 1982 Corvette, a car which, like its early forerunners, came only with an automatic transmission. After a one-year drought, a four-speed manual returned to its proper place on the options list when the redesigned fourth-generation Corvette appeared as a 1984 model. A true fuelie Corvette returned in 1985 with the arrival of Bosch's tuned-port injection setup,

which carried over the sensational 300 horsepower, 5.7-liter, second-generation LT1 V-8, introduced in 1992. The TPI LT1 remained the Corvette's heart and soul until replaced by the sequential-port fuel-injected LT1 in 1994. It was then superseded by the thoroughly modern LS1 in 1997.

Honors in recent decades include five trips around the legendary Brickyard in Indianapolis as the prestigious pace car for the Indy 500. In all five years (1978, 1986, 1995, 1998, and 2003), street-going replicas of those official Indy pacers were sold to the public. Special high-profile models were also created to mark the Corvette's 25th, 35th, 40th, and

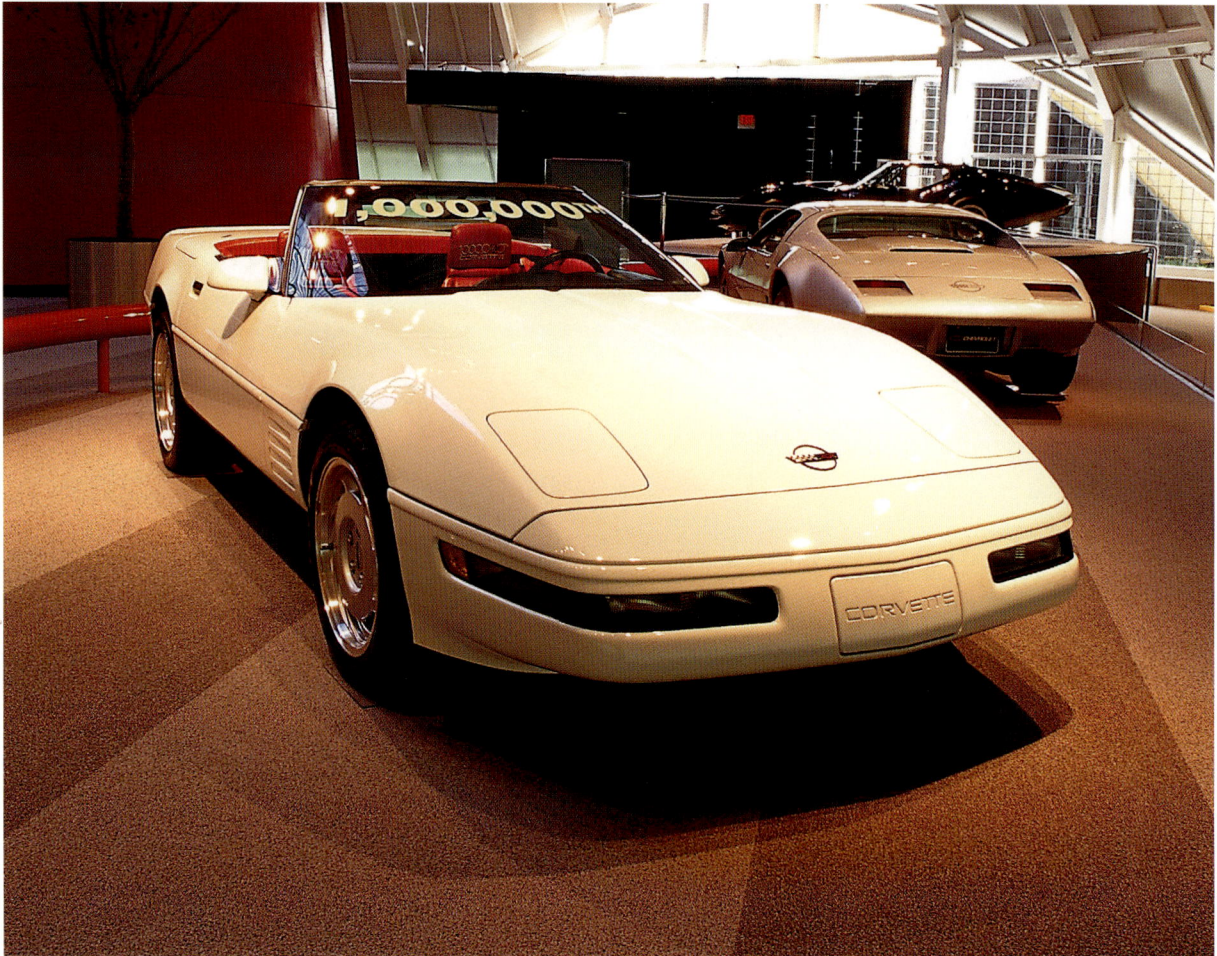

On July 2, 1992, the Bowling Green assembly plant in Kentucky rolled out the celebrated one millionth Corvette, which was a white convertible with red interior. Chevrolet later donated that car to Bowling Green's National Corvette Museum, where it resides today.

50th birthdays in 1978, 1988, 1993, and 2003, respectively. Equally special collector editions were offered in 1982 and 1996. The Corvette Grand Sport, a modern-day, regular-production commemoration of its all-out race-ready namesake of 1963, was also new for 1996.

Easily one of the greatest moments in Corvette history came in 1989 when the feared and revered ZR-1 debuted as a 1990 model. The ZR-1's high-tech aluminum LT5 V-8, 375 horsepower, 5.7-liter small-block with dual overhead cams, quickly helped remind many by-standers of some earlier exotic Corvettes. Prime examples include the 425 horse-power LS-6 of 1971, the all-aluminum ZL-1 of 1969, and the aluminum-head L88s built from 1967 to 1969. Unlike those grumpy, race-bred, ultra-low-production beasts, the ZR-1 was an animal that was easy enough to get along with on the street, yet more than capable of ripping the lungs out of any and all stoplight challengers. It also stuck around long enough

Easily recognized by its widened tail section, which was added to help house some serious rubber in back, Chevrolet's first ZR-1 Corvette emerged in 1990 to battle with the world's best sports car. With a top end of nearly 180 miles per hour, the 375 horsepower LT5-powered ZR-1 was an able competitor. LT5 output jumped to 405 horsepower in 1993. ZR-1 production ended two years later.

Improvements in 1994 made the LT1 feel stronger even though the advertised output remained at 300 horses. Modifications included trading the previously used multi-fuel injection for a sequential-port setup. Speed-density fuel calibration was also superseded by a more precise mass-air flow calibration system.

Beneath the 1993 40th Anniversary Corvette's Ruby Red Metallic skin was the heart of the exceptional 5.7-liter LT1 small-block V-8, a 300 horsepower thriller that was first introduced in 1992. By 1994, the LT1 was the backbone of Chevrolet's performance lineup and powered everything from the Z28 Camaro to the Impala SS.

The sixth-generation Corvette was unveiled to the public in January 2004.

to make a noticeable impression, although market pressures brought about the ZR-1's demise in 1995.

Even without the ZR-1 to lead the way, Chevrolet's Corvette remained a force to be reckoned with, even more so after the LS1-powered C5 model debuted for 1997. Those who mourned the ZR-1's passing were reenergized in 2001 when the reincar-nated Z06 Corvette appeared with its reborn LS6 V-8. Now we're all anxiously waiting to see just how hot the new C6 will be.

Five decades down the road and America's sports car is still out there running strong, looking hot, and proving it all night. How many of us today wish we could say the same?

The Corvette's groundbreaking status as America's first true mainstream sports car is enough to forever honor those 300 Polo White 1953 roadsters as milestones in American automotive history. The fact that its builder was Chevrolet—a leader in frugal practicality—only made the Corvette's birth even more historic.

Before 1953, the low-priced field was the last place a speed-conscious customer looked for performance, let alone sports car performance. At the time, sports cars were reasonably popular in America, and most were imported from Europe. All previous attempts to market Yankee reactions to this European postwar invasion had basically consisted of low-production independent efforts or special hybrids made up of American engines in foreign bodies. Chevy's entry into this field, however, was red, white, and blue through and through.

Although much of the credit for the Corvette's development falls on the shoulders of longtime chief engineer Zora Arkus-Duntov, he didn't work for General Motors when the project began. By the time he joined Chevrolet's research and development team in May 1953, the wheels were already turning and initial production startup was one month away. While Duntov soon boldly made his presence known, it was actually Harley Earl, legendary GM styling head from 1927 to 1958, who may represent the true "father of the Corvette."

Earl began toying with the idea of a sporty regular production two-seater in the fall of 1951. A plaster model was ready by April 1952, and support for the project came the following month from Chevrolet chief engineer Ed Cole. Chevy engineers were brought into the fray in June and kicked off a mad rush to build a showcar prototype for GM's upcoming Motorama at New York's Waldorf-Astoria Hotel in January 1953.

Under Earl's direction, designer Robert McLean laid out the basic platform on a 102-inch wheelbase. The chassis was a mixture of stock Chevy parts (front suspension), specially adapted off-the-shelf components (steering and rear axle), and newly designed pieces (the rigid X-member frame). A definitely unique fiberglass shell went on top, which was supplied by the Molded Fiber Glass Body Company of Ashtabula, Ohio. Although the original Corvette image was quickly dated as the style-conscious 1950s marched on, it was well-received in 1953, since it was considered both modern and sporty.

For power, Cole's engineers tweaked Chevy's 235-ci six-cylinder up to 150 horsepower from the

All 300 1953 Corvettes were painted Polo White and had red interiors. Like many of its sports car rivals from Europe, the first Corvettes didn't have exterior door handles or roll-up windows.

Along with the painful fact that all 1953-54 Corvette's used Chevrolet's mundane two-speed Powerglide automatic transmission, customers also found fault with the location of the tachometer, hidden behind the steering wheel in the center of the dash. Most agreed the passenger had a better view of rpm readout than the driver.

maximum 115 horses in passenger cars. A boost in compression from 7.5:1 to 8:1, a bumpier mechanical cam, dual exhausts, and three Carter carburetors on a special side-draft aluminum manifold did the trick. A floor-shifted Powerglide automatic, renamed the Blue Flame, was behind the six and was equipment that didn't exactly impress America's fledgling sports car crowd.

According to Maurice Olley, head of research and development, "The use of an automatic transmission has been criticized by those who believe sports car enthusiasts want nothing but a four-speed crash shift. The answer is that the typical sports car enthusiast, like the 'average man,' is an imaginary quantity. Also, as the sports car appeals to a wider and wider section of the public, the center of gravity of this theoretical individual is shifting from the austerity of the pioneer towards the luxury of modern ideas. There is no need to apologize for the performance of this car with its automatic transmission," said Olley.

Others weren't so sure. "That statement should get a rise from 100,000 *Road & Track* readers," wrote *R&T's* John Bond, who nonetheless generally praised the 1953 Corvette for its performance from a ride and handling perspective. However, the automotive press pointed out that the Corvette would never be able to compete with foreign rivals on a track thanks to the Powerglide. On the street, however, it was another story. According to *Road & Track's* test, an early six-cylinder Corvette could do 0 to 60 miles per hour in 11 seconds, and the quarter-mile in 17.9 seconds—not bad at all for the time.

Generally speaking, it was a decent start for a hastily created new car to compete in a completely unfamiliar field. Nonetheless, American buyers were not entirely impressed and Corvette sales lagged well below projections through 1954, and GM officials considered killing the project. Only 700 Corvettes were built for 1955, and a decision to go forward had to be made. Hope for the future, however, had already arrived in the form of the Corvette's first V-8.

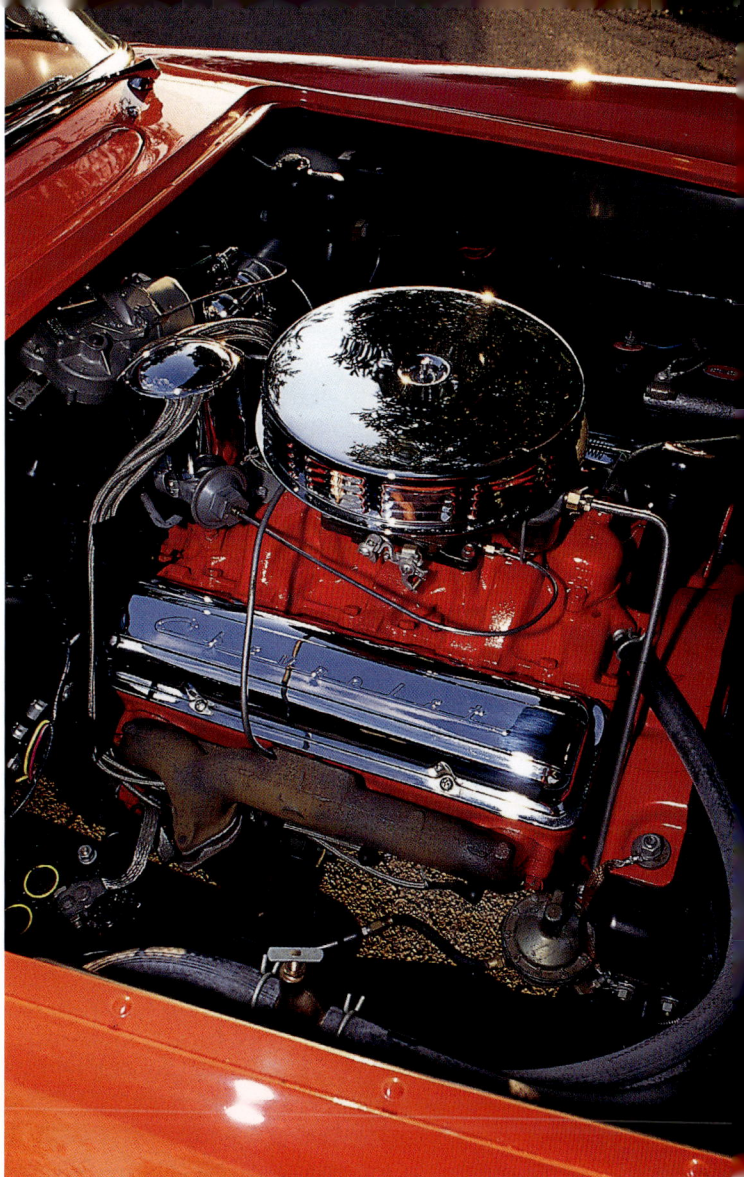

Chrome only helped sweeten the pot for V-8 Corvette buyers in 1955. With a four-barrel Power Pak, the 265 V-8 was rated at 195 horsepower beneath a fiberglass hood.

In the spring of 1954, work had begun on a V-8 Corvette, which used Chevrolet's prototype for its all-new OHV V-8 for passenger cars to debut in 1955, under the direction of Mauri Rose, head of performance development. Adding V-8 power to the package was undoubtedly the right thing to do to keep the project alive, but the appearance of a

Three sidedrift carburetors and a split exhaust manifold represented the most obvious modifications that helped transform Chevy's durable Stovebolt six into the 1953 Corvette's Blue Flame Six. Output was 150 horsepower.

plastic mockup from Ford in February 1954 that featured a "personal luxury" concept also helped. That mockup was for the legendary Thunderbird, which debuted in October 1954 with a standard V-8 beneath the long, scooped hood. Although the two-seat T-bird wasn't exactly direct competition, it was more than enough of a threat to inspire rapid-fire reaction in the Chevrolet camp.

An optional V-8 joined the Corvette lineup for 1955. At 265 cubic inches, Chevrolet's new OHV V-8 was rated at 195 horsepower beneath a fiberglass hood. A long-awaited three-speed manual transmission was also promised for 1955, but it didn't arrive until well into the model year. Although most critics still complained about the car's brakes and automatic transmission, it was clear the Corvette was on the right track to recovery, thanks to the welcomed power boost.

Saving the young legacy from extinction was the only goal in 1955, and the first V-8 Corvette helped do just that.

Additional exterior colors were added to the Corvette appeal in 1955, the same year Chevrolet's first overhead-valve V-8 appeared. While the six-cylinder was still available and up five horses that year, the majority of 1955 Corvettes featured the 265-ci V-8. Some cars late in the year were equipped with the newly offered three-speed manual transmission.

2 1956–62 Racing Improves the Breed

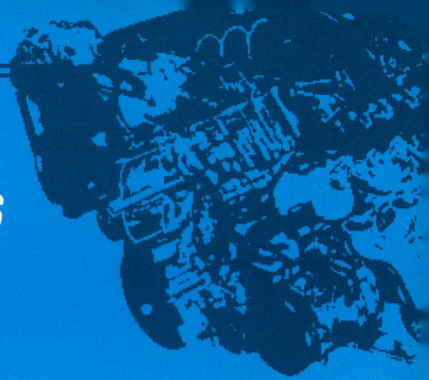

Following the 265-ci V-8's debut in 1955, an exciting new body appeared for 1956. Appeasements to Yankee sensibilities produced roll-up windows, external door handles, and an optional removable hardtop.

More importantly, at least from a horsepower hound's perspective, the 265-ci V-8 was boosted to 225 horsepower with the addition of RPO (regular production option) 469, which was made up of two Carter four-barrel carburetors. If you really wanted to get serious, you could've ordered the so-called "Duntov" cam, which was specified "for racing purposes only." Listed under RPO 449 or 448, depending on your source, the Duntov cam was only available with RPO 469. No official advertised horsepower figure was given, but most sources put output for the special-cam engine at 240 horsepower.

Now armed with a race-ready rocket, Duntov set out to prove that his baby was no longer intended just for streetside duty. In February 1956, he took a three-car team to Daytona Beach, Florida, for NASCAR's annual Speed Week trials. Veteran race driver John Fitch and champion aerobatic pilot Betty Skelton joined Duntov for Speed Week. By the time the three finally slowed down, Duntov had established a new sports car flying-mile standard of 150.533 miles per hour, while Fitch had set a two-way record at 145.543 miles per hour. Skelton managed a 137.773-mile-per-hour two-way clocking.

Fitch led a four-car effort to Sebring's 12-Hour endurance event in March with less impressive results. Even Duntov knew the Corvette wasn't ready to take on Europe's best in endurance competition. Going flat-out in speed trials was one thing; slowing and speeding up around twists and turns was an entirely different ball game that Europeans knew how to play all too well. From the beginning, Duntov knew he needed much more of a sports car to beat foreign rivals on their turf.

Nonetheless, Corvettes quickly established themselves as able-bodied competitors in Sports Car Club of America (SCCA) stock-class racing. Armed with an ever-growing arsenal of hot factory parts, such as special brakes and beefed up suspension, Dr. Dick Thompson first proved that Chevy's fiberglass two-seater could be used as a racing sports car. In 1956, he claimed his first SCCA production-class championship at the wheel of a Corvette. Thompson won again in 1957, 1962, and 1963, and added an SCCA C-Modified title in 1960.

Featuring unique aluminum trim throughout to help save weight, this Smokey Yunick-prepped 1956 Corvette was one of the three cars taken to Daytona in February 1956 for NASCAR's annual Speed Week trials. Originally driven on the sands by Betty Skelton, today the car resides in Bill Tower's noted Corvette collection. Tower also owns an SR-2 and a 1963 Grand Sport.

A heavy duty racing-type suspension and specially cooled racing brakes officially appeared as a regular production Corvette option in 1957, as did a welcomed four-speed manual transmission. Reportedly some racers were also supplied with a few oversized fuel tanks, a feature that eventually ended up on the options list in 1959, but the biggest news was the arrival of Ramjet fuel injection.

Once atop the 1957 Corvette's enlarged 283-ci small-block, Ramjet fuel injection, supplied by Rochester, boosted output to 283 horses in top form. It was the second time a Detroit V-8 reached the one-horsepower-per-cubic-inch milestone—Chrysler had offered an optional 355 horsepower 354 Hemi V-8 for its 300B luxury cruiser the year before. Although it was a bit finicky, the fuelie 283-ci was capable of powering a 1957 Corvette from 0 to 60 miles per hour in less than six seconds.

A Chevrolet team returned to Sebring in 1957, but early Corvette competition efforts weren't limited to production-class racing. The distinctive SR-2 prototype racers had emerged the year before. Conceived by Harley Earl for his son Jerry, the first

Various modification tricks used on the Betty Skelton racer included these brake-cooling ducts. Feeding outside air through vented backing plates into the drums, this setup became a regular Corvette option. These ducts were later called "elephant ears."

Bill Tower's 1956 SR-2 was the second of three built. After Harley Earl ordered the first SR-2 for his son Jerry, Bill Mitchell requested one for himself. The second SR-2's competition debut was at Daytona in 1957, where Buck Baker recorded a flying-mile time of 152 miles per hour.

Along with the optional removable hardtop, this 1957 fuel-injected Corvette also features wider 15x5.5 wheels, as identified by the small hubcaps. This fuelie is an Airbox car, which means it is equipped with special ductwork to help supply cooler, denser outside air direct access to the Rochester injection setup. Only 43 Airbox Corvettes were built for 1957.

Corvette SR-2 featured a custom fiberglass shell on a race-ready chassis. The SR-2 incorporated all the heavy-duty suspension and brake tricks Fitch had put to the test at Sebring in 1956. In Chevrolet engineering lingo, the beefed up parts were known as SR components, which may or may not have stood for "Sports Racing" or "Sebring Racer." Either way, Fitch's Corvette factory racers were the first SRs, which meant any variation to follow was naturally the second, thus the SR-2.

Jerry Earl's SR-2—shop order number 90090—was built in about four weeks, which led some to believe its modified body was dropped onto an existing Sebring chassis. Supposedly, an off-the-lot 1956 Corvette went into Engineering in May and came out as the SR-2 in June.

A metallic blue body with an extended snout and bright aluminum bodyside cove panels was on top of that race-ready chassis. Louvers on the hood and vents in each door were functional; the former cooled the engine, and the latter cooled the rear brakes. Twin short windscreens were used up front, and a small tailfin was added down the center of the decklid in back. Racing modifications included cutout exhausts, an oversized fuel tank, and Halibrand knock-off mag wheels.

After a disappointing debut at Wisconsin's Elkhart Lake in June 1956, Earl's SR-2 was sold to Jim Jeffords, driver for Chicago dealer Nickey Chevrolet, in 1957. By that time, the first SR-2 featured a larger, reshaped tailfin offset to the driver's side to serve as a headrest and rollbar. The new design was added at Earl's request after he saw it on the second SR-2 that was built for Bill Mitchell. As for the first SR-2, Jeffords' raced it to a SCCA B/Production championship in 1958.

Mitchell's red and white SR-2 debuted in February 1957 during Daytona's Speed Week trials. There, Buck Baker recorded a flying-mile speed of 152.886 miles per hour in the high-finned car, which finished 16th at Sebring a month later.

A third low-finned SR-2 was built for GM president Harlow Curtice. Basically a stock 1956 Corvette underneath, Curtice's SR-2 was only meant for the show circuit. Metallic blue paint was used again, as were Dayton wire wheels and a removable stainless steel hardtop. All three SR-2s still survive and have passed through various owners' hands and experienced a modification or two along the way.

While SR-2 racers represented independent efforts, the next great Corvette competition project

The SR-2 was lightened throughout and featured many modifications that may or may not have been retrofitted, including a four-speed manual transmission that didn't become a Corvette option until 1957. Notice the twin racing windscreens and the cooling louvers added to the hood.

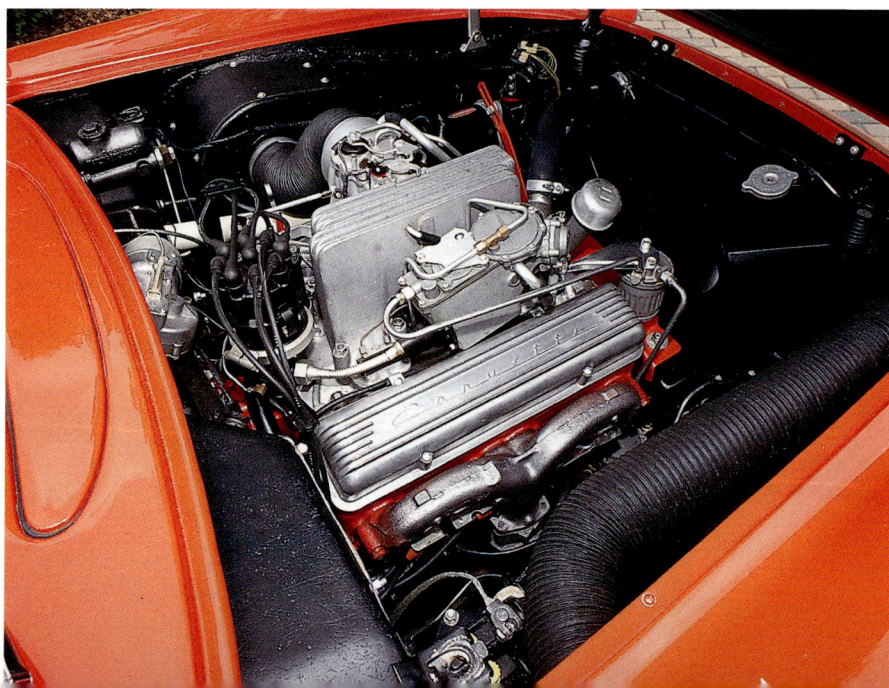

The Airbox arrangement's cool-air plenum can be seen at the top of this photo directly inside the fender—it's connected by that flexible duct to the Rochester unit. Listed under RPO 579E, the Airbox 283 V-8 was rated 283 horsepower, the same as the top 1957 fuelie small-block.

Underhood crowding created by the Airbox ductwork meant the tachometer drive had to be relocated, which meant the tach itself had to be moved from its typical spot in the dash to atop the steering column. The column-mounted tach is one of the easiest clues to the identity of an Airbox Corvette. Notice the medallion located in the opening that was left behind by the remounted tach.

was a factory job from top to bottom. Zora Duntov's fabled SS, designated XP-64, was hastily created beginning in July 1956 for competition at Sebring in March 1957. It was an all-out, purpose-built machine, with its low-slung tubular space frame and light-weight magnesium body. Aluminum was used throughout, including the gearbox, radiator, and cylinder heads on the modified 283 fuelie V-8. Coilover shocks were at all four corners, an independent de Dion rearend was in back, and the brakes were finned drums with the rear pair mounted

inboard on the differential to reduce unsprung weight. An innovative, servo-controlled booster system was also incorporated to help prevent rear-wheel brake lockup. Along with the beautiful blue SS built in Chevy Engineering, Duntov's crew also fashioned a crude fiberglass-bodied test mule counterpart.

On paper, the SS project certainly looked promising. Early tests of the white SS mule at Sebring were impressive. However, problems appeared almost immediately on race day for the blue SS and forced an early retirement from the 12-hour

29

Truly innovative throughout, the magnesium-bodied SS racer of 1957 nonetheless fell victim to hasty development. Even though it possessed the power to compete with Europe's best, the car carried too many gremlins with it onto the track at Sebring in March 1957. It retired after only 23 laps.

A modified fuel-injected small-block with aluminum heads and a magnesium oil pan powered the SS. Additional tweaks included revised valves and a special injection manifold. Dyno tests in 1957 put output at 307 horsepower.

endurance event. Any hopes of going back to the drawing board and trying again at Le Mans later that summer were dashed once the Automobile Manufacturers Association (AMA) issued its so-called ban on factory racing involvement. One can only wonder what might have been, especially after Duntov's SS hit 183 miles per hour at GM's Phoenix proving grounds in December 1958. The car was later donated to the Indianapolis Motor Speedway Hall of Fame Museum where it still resides today.

As for the fiberglass SS mule, it also survives, but not in its original form. During the winter of 1958-59, it was acquired by Bill Mitchell, head of GM styling. Its chassis was used as a base for his XP-87 Stingray racer, a car that provided more than one styling trick later used on the regular production Sting Ray in 1963. With Dr. Dick Thompson at the wheel, Mitchell's Stingray roared to an SCCA C/Modified championship in 1960.

After the AMA racing ban, Chevrolet was forced to cut back on its factory-backed competition efforts. While support of certain Corvette racers—namely Dr. Thompson—continued, it was primarily the covert, back-door type of support. The long list of factory

Veteran driver John Fitch and co-driver Piero Taruffi sat here on race day at Sebring in 1957 after Stirling Moss and Juan Fangio withdrew from the SS team prior to the event. One of the SS racer's many glitches involved heat buildup inside the cockpit. Fitch and Taruffi literally boiled as the magnesium shell didn't dissipate heat like its fiberglass counterpart.

Penned by stylist Larry Shinoda, the XP-755 Mako Shark was built in 1961 as a personal car for Shinoda's boss, Bill Mitchell. The XP-755 borrowed many of its lines from the XP-720 project, which was the prototype for the all-new 1963 Sting Ray. When another Mako Shark styling exercise appeared in 1966, Mitchell's car became known as the Mako Shark I. Today, the Mako Shark I resides at the National Corvette Museum in Bowling Green, Kentucky.

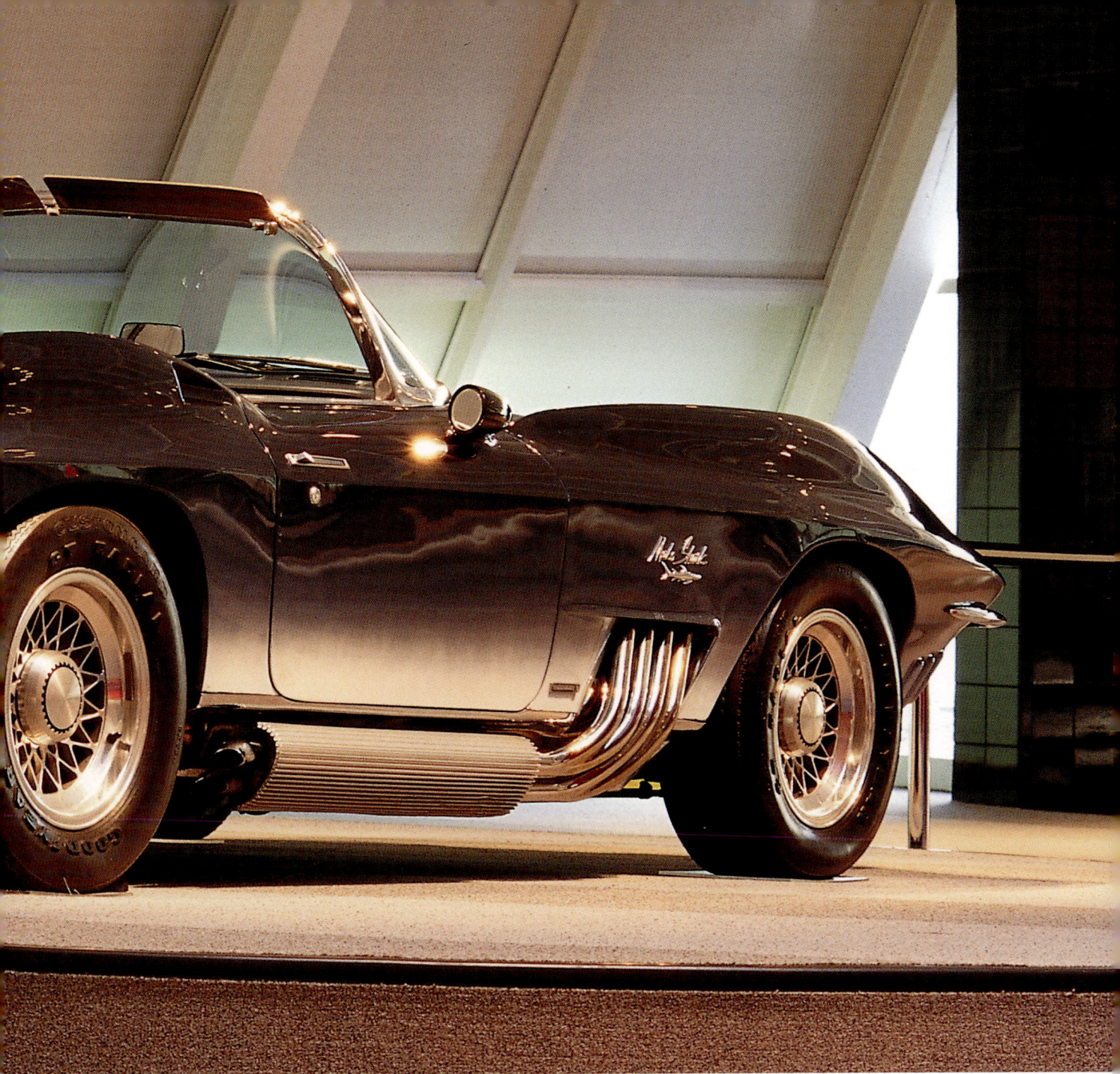

This 1961 fuelie features various racing-inspired options, including wide wheels, quicker steering, and heavy-duty suspension with specially cooled brakes. The rarest of the bunch is LPO (limited production option) 1625, the big tank. Notice the exposed fuel filler cap behind the door—it represented one of the modifications made to mount the larger tank behind this 1961 Corvette's seat.

Adding the oversized 24-gallon fuel tank meant the removable hardtop option was also required since the tank took up the space behind the seat where the folding top normally resided. The oversized tank was first officially offered for 1959, although some reportedly had been supplied to racers as early as 1957.

Chevy's small-block displaced 265 cubic inches when it was introduced in 1955, then grew to 283 cubic inches in 1957. In 1962, the small-block increased to 327 cubic inches. The 327 for 1962, rated at 340 horsepower, is shown here.

options was still present and accounted for after 1957: big metallic brakes with special cooling ducts, quicker steering gear, bigger wheels, and beefed-up suspension components. Duntov even tested a set of weight-saving aluminum fuelie cylinder heads in 1960, although that option was quickly discontinued

when production defects couldn't be cured. All this purposeful equipment and more was on the option list, as were loads of fuel-injected power—up to 360 horses worth by 1962.

American sports car devotees who thought it just couldn't get any better only had to wait another year.

From nose to tail, the all-new 1963 Sting Ray was a stunner. It may easily rank among the most startling transformations in American automotive history. "This is the one we've been waiting for," wrote *Motor Trend's* Roger Huntington. "This is a modern sports car." Even Duntov himself was finally satisfied: "For the first time I now have a Corvette I can be proud to drive in Europe."

The sleek Sting Ray was nonetheless more comfortable inside than its solid-axle forerunner, thanks to repositioned seating in a redesigned frame. It was lower, thinner, shorter, and rode on a compact 98-inch wheelbase (down from the previous 102-inch chassis). Innovative (at least in Yankee terms), independent rear suspension improved both ride and handling.

On top of it all was Mitchell's alarming new Corvette shell—available for the first time in coupe form—with hideaway headlights in front and crisp, curvaceous lines throughout. It featured a tapered roofline with Mitchell's pet split-window stinger theme in back. Any way you looked at it, the car was a killer, although many critics—including Duntov—didn't think much of the split rear windows. Despite Mitchell's adamant defenses, the stinger was deleted in 1964.

Initial demand for the attractive Sting Ray was so great that an extra shift at the St. Louis plant couldn't even help keep up. Overall, 1963 sales soared by 50 percent to a new high of 21,513 cars. Still a polite tourer in base form, the 1963 Sting Ray typically could've been transformed into a street killer, thanks to a long list of options. The most prominent option, from a power perspective, was the 360 horsepower fuel-injected small-block, RPO L84.

Last, but certainly not least, on the options list was RPO Z06, the Special Performance Equipment group. RPO Z06 consisted of every hot part on the Corvette shelf. Mandatory features included the L84 fuelie 327, backed by its close-ratio Muncie four-speed and Positraction rearend. Heavy-duty suspension parts, special "cerametallix" power brakes with unique cooling features, an oversized 36.5-gallon fiberglass fuel tank, and five cast-aluminum knock-off wheels were initially listed. In December, Chevrolet temporarily canceled the knock-off option due to production difficulties and removed the big tank from the package to help whittle down RPO Z06's original $1,818.45 asking price. The 36.5-gallon tank, RPO N03, remained a separate option that could be added to any Sting Ray coupe.

Zora Arkus-Duntov tried again to build a world-class racing Corvette late in 1962. Plans to build 125 lightweight Grand Sports failed when GM's anti-performance overlords shot things down. Only five Grand Sports were built before the ax fell early in 1963. Two of these were later converted into roadsters.

Beefy brakes were part of the Z06 deal. Included was a dual-circuit power booster; enlarged, finned drums; sintered cerametallix linings; and special cooling gears. Rubberized elephant ear ducts directed airflow through vented backing plates where an internal fan helped stir things around. The drums themselves were also vented. One drawback to these brutish binders was that they barely worked at all before they were warmed up.

Reportedly, only 199 buyers chose a Z06 Sting Ray coupe in 1963.

The Z06 Sting Ray wasn't the meanest Corvette built for 1963. That honor went to the Grand Sport, a purpose-built race car reminiscent of Duntov's ill-fated SS of 1957. Even though the 1957 AMA racing ban had supposedly closed the door on such shenanigans, Duntov never stopped thinking about building a world-class competition Corvette. Once performance-conscious Semon E. "Bunkie" Knudsen came over from Pontiac to become Chevrolet's general manager in November 1961, Duntov was allowed just enough leeway to kick off another competition-minded project.

The competition project began in the summer of 1962, as Duntov's engineers began fashioning a special lightweight Corvette, based on a tubular-steel, ladder-type frame. Weight was also saved by using various aluminum components, Halibrand magnesium knock-off wheels, and a special hand-made fiberglass body with super-thin panels. Some external dimensions were also changed to help improve the stock Sting Ray's aerodynamics. Plexiglass windows and a 36.5-gallon fuel tank were

Grand Sport racers were amazingly stock looking inside, but a closer inspection revealed the presence of a 200-mile-per-hour speedometer. Barely noticeable just above the passenger seat is a movie camera used by Chevrolet engineers to document on-track testing action.

Beneath all that plumbing is a 377-ci aluminum small-block. Those are four 58mm Weber side-draft two-barrels on a cross-ram intake. The Webers on the left feed the cylinder bank on the right and vice versa. Original plans to equip the Grand Sport with the 377-ci small-block were canceled when the project was shut down by GM's front office in January 1963. Grand Sports were first powered by less exotic fuel-injected 327s. Various other power sources followed over the years.

installed, and the brakes were large 11.75-inch Girling discs.

Initial specifications called for a 377-ci small-block fed by four Weber carbs. Early plans also mentioned a production run of 125 or so Grand Sports. Neither became reality. The first Grand Sport was fitted with an aluminum 327 fuelie while the 377-ci V-8 was still in development. Before that work could be finished, GM announced a ban of its own and instructed all divisions in January 1963 to cease racing projects immediately.

Only five Grand Sports escaped Chevrolet Engineering before that order came down. From

there, each went through a steady progression of independent race teams. They also underwent various mechanical and exterior modifications and took on a varied succession of scoops, flares, and engines. Both small- and big-blocks were used over the years, and two of the coupes were later converted into roadsters for competition at Daytona in February 1964.

GM officials stepped in again before those two roadsters could reach Florida. Chevrolet Engineering was still supporting much of the Grand Sport racing effort, despite orders to the contrary a year before. This time, GM executives instructed

While a host of racing options had been available to Corvette buyers since 1957, Chevrolet really got the ball rolling in 1963 and grouped a host of parts together in one package, RPO 206. Only 199 206 Sting Rays were built for 1963.

A performance powerplant from head to toe, the L78 396 pumped out 425 rompin', stompin' horses. Mandatory options included the close-ratio M20 Muncie four-speed, transistorized ignition (RPO K66), and a Positraction differential. Notice the optional power brakes with dual-circuit master cylinder.

Knudsen to end these activities or risk his annual bonus. The five Grand Sports were then sold off. The last major competition appearance was in 1966. Like the SS and SR-2s before them, all five Grand Sports survive today in collectors' hands.

Back on the street, another milestone moment in Corvette history came in 1965, when four-wheel disc brakes were made standard equipment and two great powerplants crossed paths. Chevrolet's fuel-injected small-block—a major part of the Corvette mystique and the top power source since 1957—was offered for the last time in 1965. Its $538 ask-ing price was no longer justifiable, as less expensive, less finicky carbureted 327s had become nearly as powerful. Even more powerful was the all-new 396-ci Mk IV V-8, the Corvette's first big-block.

Introduced early in 1965, the 396 featured a bulletproof block and free-breathing cylinder heads with ball-stud rockers and large, canted valves. These exceptional heads, in concert with a high-lift mechanical cam, 11:1 compression, transistorized ignition, a big Holley four-barrel on an aluminum intake, and header-type cast-iron exhausts, helped the Mk IV big-block produce 425

A distinctive triangular air cleaner was part of the L71 package. Under normal operation, the 435 horsepower 427 was fed by the middle Holley two-barrel. Putting the pedal to the metal brought the other two carbs into play via a vacuum signal. When really wailing, the 3x2 setup sucked in some 1,000 cfm worth of fuel/air.

horsepower. The following year, the Mk IV was bored out to 427 cubic inches. Then in 1967, three Holley two-barrel carburetors were added on top to help produce 435 horsepower.

The optional 3x2 equipment was one of the best working progressive throttle arrangements ever seen on an American multiple-carb setup, and was both efficient and hot to trot. Under normal driving conditions, the middle two-barrel worked alone and pumped roughly 300 cfm of fuel/air into the lazily loping L71 big-block. When the hammer dropped, a vacuum signal brought the other two carbs into action, bringing total flow to about 1,000 cfm as all hell broke loose. In the words of *Hot Rod's*

Eric Dahlquist, the tri-carb Sting Ray was perhaps the "hottest 'Vette yet."

Actually, the hottest Corvette in 1967 was another racing-inspired Sting Ray, the legendary L88. Far more cantankerous than the 1963 Z06, the 1967 L88 Corvette was clearly not meant for street use, but some drivers tried. While Chevrolet gave the L88 427 a token rating of 430 horsepower, the actual output was probably upwards of 550 horses. Aluminum heads with large valves were part of the L88 deal, as were 12.5:1 pistons and a huge 850-cfm Holley four-barrel that fed cooler, denser air by a specially ducted hood. Mandatory options included transistorized ignition, power-assisted metallic

After big-block displacement was increased to 427 cubic inches in 1966, the Corvette's top performance V-8 was bumped up to 435 horsepower in 1967, thanks to the addition of three Holley two-barrel carburetors. Listed under RPO L71, the tri-carb 427 found 3,754 buyers in 1967.

Beneath the 1967 L88's functional hood was the star of the show, the aluminum-head 427. The odd-looking air cleaner fit into special ductwork in the hood's underside from where it drew denser air from the base of the windshield. Notice the black road draft tube running from the driver's side valve cover to behind the master cylinder. An L88 couldn't pass emissions tests because it vented its crankcase directly into the atmosphere.

brakes, F41 sports suspension, Positraction, and the indestructible M22 Rocker Crusher four-speed. The RPO C48, the heater-defroster delete, was also included with the L88 package. Who needed such luxuries on a race track, right?

On the legendary track at Le Mans in France, one of the 20 L88 Corvettes built for 1967 impressed all with its dominating speed down the Mulsanne Straight. A thrown rod halfway through the race ended yet another Corvette attempt at international racing glory. Not all was lost, however, as the aluminum-head L88 returned in 1968, and became a big SCCA winner. L88 production was 80 in 1968, followed by another 116 in 1969.

Although the second-generation Corvettes, the mid-year models of 1963-67, never produced the world-beater Duntov had hoped for, they didn't dim the original Sting Ray's reputation on the street in the least. They were all great cars.

Owned by former GM engineer Bill Tower, this L88 Corvette may be the first of the 20 built for 1967. It features various non-stock features including prototype Rally Wheels and L-88 decals. The racing side-pipes were used during proving ground tests.

Less expensive than fuel injection and considerably more powerful, the 396-ci L78 V-8 debuted as the Corvette's first big-block in 1965. There were 2,157 396 Sting Rays made that year.

Work on a restyled, modernized Sting Ray began early in 1965 with high hopes of making this new ideal ready for production in 1967. Development problems delayed the debut and forced Duntov, Mitchell, and the rest to roll out one more mid-year model before the third generation finally bowed for 1968.

Inspired by stylist Larry Shinoda's Mako Shark II showcar of 1965, the 1968 Corvette was quickly compared to a Coke bottle with its bulging front and rear quarters that served as bookends for a slimmed-down mid-section. Comparing the new 1968 to its "pinched-waist" 1963-67 predecessor was akin to standing Raquel Welch next to Lily Tomlin. Along with being seven inches longer and considerably more shapely than its forerunner, the 1968 Corvette also scored higher in the sultry department.

As for the stuff of legends, top power for the lusty 1968 Corvette once more came from the 427 big-block, again available in optional 435 horsepower L71 3x2 form. The race-ready L88s were carryovers for both 1968 and 1969. Those who wanted a piece of the L88 lightweight action without the off-road limitations could add the L89 aluminum head option to their L71 427. The $832 L89 option was first offered in 1967 and attracted 624 buyers in 1968, after only 20 pairs were

sold the previous year. Another 390 pairs went out the door in 1969. Although valves and combustion chambers differed slightly for the lightweight L89 heads compared to their cast-iron counterparts, no change in advertised output was made when they were added to the 435 horsepower 427.

Even more aluminum was used in 1969, when Chevrolet unleashed two ZL-1 Sting Ray coupes, the last of the truly exotic Corvettes to make it to the streets. Featuring a pair of aluminum heads atop an aluminum cylinder block, the ZL-1 427 was not for the timid, of spirit or wallet. Sticker price for RPO ZL-1 alone went well beyond four grand. The ZL-1 was much more radical than the L88 throughout, but it still carried the same token horsepower rating—430 horsepower. Actual output soared past 500 horses, a fact that was quickly proven at the track where a 1969 ZL-1 Corvette wowed the press with a screaming 12.1-second quarter-mile pass.

Plans for yet another high-powered aluminum-aided big-block for 1970 ran afoul of Chevrolet's latest "deproliferation" plans. Originally listed in 1970 Corvette paperwork and tested by the press in prototype form, Chevy's stillborn LS-7 454-ci big-block featured aluminum heads, 12.25:1 compression, and 465 horsepower—more than enough to

Chevrolet's Mk IV V-8 ballooned to 454 cubic inches by the time the last big-block Corvette was offered in 1974.

Chevrolet's third-generation Corvette received an all-new Coke-bottle body for 1968. Removable roof panels were also new that year. This 1968 coupe is powered by an aluminum-head L89 427. Priced at $805.75, the L89 option was checked off 624 times in 1968.

Originally introduced for the tri-carb L71 427 in 1967, the L89 aluminum-head option shaved off some unwanted pounds, and added a revised combustion chamber and different valves. Advertised output, however, remained at 435 horsepower. PRO L89 was discontinued after the last 427 was offered in 1969.

put an LS7 Sting Ray into the 13-second quarter-mile bracket, according to *Sports Car Graphic*. Growing corporate concerns over how much power was too much, led GM's front office to kill the LS-7 before it made regular production, leaving Corvette buyers to make do with the much more mundane 390 horsepower LS5 454.

Even with Detroit's horsepower race rounding its last turn, Chevy engineers managed one last gasp in 1971 when they made the LS6 454 a Corvette option. LS-6 features included open-chamber aluminum heads, a big dual-feed Holley four-barrel atop an aluminum intake, transistorized ignition,

and a mechanical cam. Advertised output was 425 horsepower. Quarter-mile performance was listed at 13.8 seconds by *Car and Driver*.

Big-block Corvettes weren't the only headline makers in the 1970s. Chevrolet's long-running small-block was enlarged again in 1969 to 350 cubic inches. In 1970, the 350 was the base for the Corvette's hottest small-block since the 375 horsepower L84 327 fuelie had disappeared after 1965. Initially rated at five less horses than the L84, the LT1 350 made for a better-balanced, more road-worthy Corvette compared to the nose-heavy Mk IV models. The LT1 line-up included a solid-lifter cam, 11:1 compression, and

The sexy, new 1968 Corvette measured seven inches longer than its 1967 forerunner and was nearly two inches shorter in height. Base big-block power came from a 390 horsepower 427, as this Silverstone Silver coupe demonstrates.

Except for the air pump emission controls, the 1969 ZL 1 427 looks very much like the L88. Unlike the L88, the ZL 1 has an aluminum cylinder block to accompany those aluminum heads. Chevrolet also used the all-aluminum ZL1 427 in Camaros for 1969.

a Holley four-barrel on an aluminum intake. After dropping to 330 horsepower in 1971, the last of the first-generation LT1s appeared in 1972 with a net rating of 255 horses.

Chevrolet also offered the ZR-1 package, an LT1 Corvette with an M-22 four-speed, heavy-duty power brakes, special suspension, and an aluminum radiator. RPO ZR-1 was listed each year along with the LT1 and cost around $1,000. Production was only 25 in 1970, 8 in 1971, and 20 in 1972. A similar equipment group, RPO ZR-2, was available along with the LS6 big-block in 1971. Priced at $1,747, ZR-2 equipment found its way into a mere eight 425 horsepower Sting Rays that year.

After 1971, much of the Corvette's serious sting fell by the wayside as Detroit's era of outlandish

Identified by its nose stripe and bulging hood, this ZL 1 Corvette is one of only two built for 1969, although others may have resulted from crated engines being delivered into private hands.

Chevrolet's original LS6 V-8 displaced 454 cubic inches and was rated at 425 horsepower. It was only offered under a Corvette hood for one year, and 188 LS6 Corvettes were built in 1971.

Air conditioning typically wasn't available along with RPO LT1 when the option debuted in 1970. Per Zora Arkus-Duntov's orders, one 1972 LT-1 was taken off the line and tested with an air-conditioning installation and resulted in the availability of air-conditioned LT1 Corvettes by the end of the year. The 1972 LT1 shown here is that very prototype.

performance came to a close. Drastically lowered compression ratios in 1971 were followed by net-rated output figures the following year. Along with the 255 horsepower LT1, speed-conscious customers could also pick the 270 horsepower 454 in 1972. The Mk IV V-8 wasn't long for the world, either. Even though the Corvette's third-generation ran up through 1982, the end of an era came in 1974 when Chevrolet built its last big-block Corvette. For Corvette buyers from then on, it was a small-block or no block at all.

Chevrolet's first-generation LT-1 Corvette was offered between 1970 and 1972 as a nimble small-block alternative to those big-block bullies. This 1972 LT-1 is one of 1,741 built.

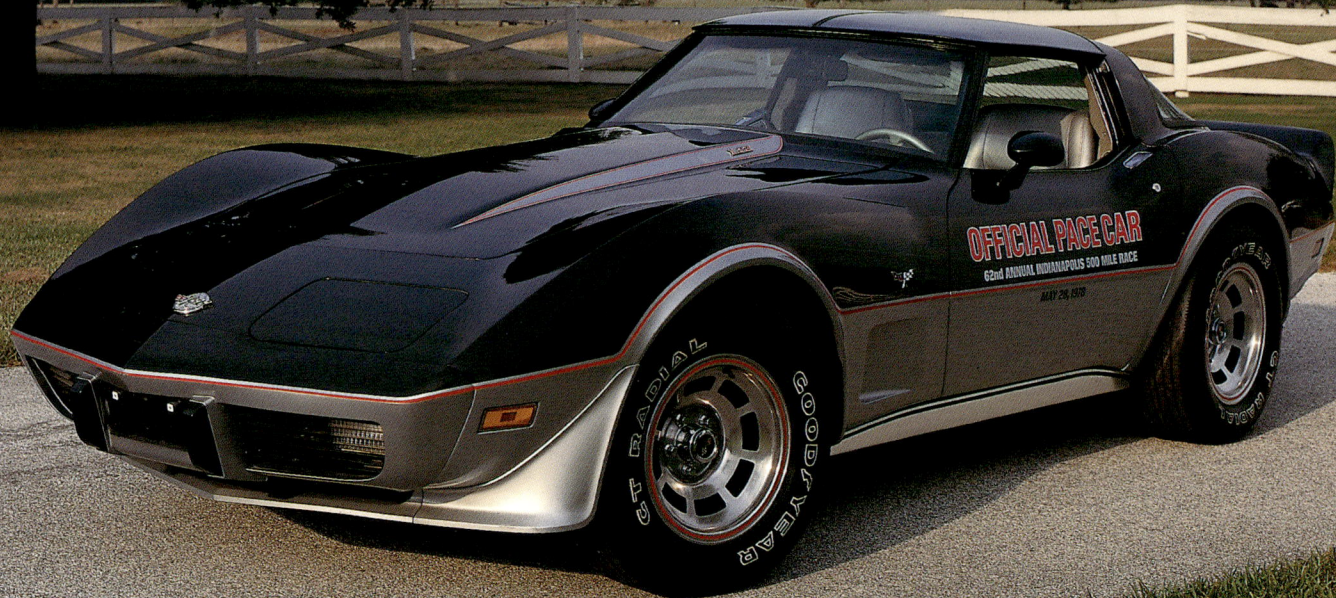

5

1975–96
Chevrolet's Fiberglass Legacy Rolls On

Let's face it, the late 1970s were simply bad years as far as Detroit performance was concerned, and America's only sports car was no exception. As if sky-high insurance costs and ever-growing safety concerns weren't enough to discourage the building of high-powered automobiles, emissions standards strangled the life out of the good ol' Yankee V-8. Once the doomed 454 big-block disappeared from the Corvette options list after 1974, drivers were left with a series of continually weakening 350 small-blocks.

The situation started to turn around by 1978, the year Chevrolet celebrated the Corvette's 25th birthday. Thanks to the optional 220 horsepower L82 350, the 1978 Corvette was still king of the American hill, a fact that was not missed by *Car and Driver*. "We can happily report the 25th example of the Corvette is much improved across the board. Not only will it run faster now—the L82 version with four-speed is certainly the fastest American production car—but the general drive-ability and road manners are of a high order as well."

Along with that welcomed L82 shot in the arm, all 1978 Corvettes received special 25th anniversary badges to mark the special occasion. The 25th Corvette also wore a new fastback rear window that aided rearward vision and improved storage space behind the seats.

Additional commemoration was initially available through RPO B2Z, which added an exclusive two-tone silver anniversary paint scheme. Dual sport mirrors and aluminum wheels were required options, along with the paint. There were 15,283 silver anniversary Corvettes produced.

A special Limited Edition 1978 model marked the Corvette's 25th anniversary, as well as the Corvette's first appearance as the pace car for the Indy 500. The Limited Edition Indy Pace Car replica, priced at $13,653.21, was stuffed full of options that included power windows, door locks, and antenna; removable glass roof panels; a rear window defogger; air conditioning; sport mirrors; and a tilt-telescopic steering column. Other options included white-letter P225/60R15 tires, a heavy-duty battery, and an AM/FM 8-track stereo with dual rear speakers. A front air dam, rear spoiler, and aluminum wheels with red pinstripes completed the deal, which was originally intended to be a limited status, but ended up quite the contrary. When the feeding frenzy for what many felt would be a future collectible came to a close, Chevrolet sold 6,502 of these high-profile fastbacks. Today, their collector value remains minimal.

Another special-edition model came four years later as the third-generation Corvette bowed out.

Chevrolet marked its first Indy Pace Car appearance with this Limited Edition model in 1978. Among a long list of features were the front air dam, aluminum wheels, and rear spoiler. All 1978 Corvettes were 25th anniversary models.

Chevrolet temporarily shelved its topless Corvette in 1975. A convertible model returned in 1986.

The 1982 Collector's Edition featured a unique paint scheme, this time a silver-beige finish accented with graduated grey decals and pinstripes. Features included hatchback rear glass, special emblems, exclusive turbine alloy wheels with white-letter P255/60R15 rubber, a leather-wrapped steering wheel, matching silver-beige leather upholstery, and luxury carpeting. Removable glass roof panels with a unique bronze tint, a rear-window defogger, and a power antenna were also included in the package. The price was about $22,500, which was more than $4,000 beyond the base coupe's sticker. In chief engineer Dave McLellan's words, the 1982 Collector's Edition was "a unique combination of color, equipment, and innovation to produce one of the most comprehensive packages ever offered to the Corvette buyer." Production was 6,759.

Window dressing aside, real history was made in March 1983 when Chevrolet finally introduced the all-new Corvette. The third-generation Corvette was a decade old when Dave McLellan began envisioning a redesigned next generation. Initially, it appeared his vision would become reality in 1983,

Left: As part of a charity fundraising effort, the National Council of Corvette Clubs raffled off the very first, serial number 00001, of the all-new 1984 Corvettes. Today, the car is in collector Dick Gonyer's hands and is still identified on the doors and windshield as it was when originally raffled. This is the only #00001 model from any Corvette generation known to survive.

Below: Chevrolet celebrated an important Corvette birthday for the second time in 1988 and offered a 35th Anniversary Edition. Production was 2,050 vehicles.

A Corvette convertible returned in 1986 just in time to become the second fiberglass two-seater from Chevrolet to pace the Indianapolis 500. This 1986 Indy Pace Car replica resides in the Klassix Auto Museum in Daytona Beach, Florida.

They called it the King of the Hill for good reason. Chevrolet's first ZR-1 was easily the most dominating street-going Corvette ever built. With a superb suspension, loads of rubber, and 375 horses beneath its clamshell hood, the 1990 ZR-1 could run with anything in this country, and almost anything worldwide.

Like its forefather 35 years earlier, all 1988 35th anniversary Corvettes were white. This external identification was included in the $4,796 package listed under RPO Z01.

from engine to differential as one rigid component joined by an aluminum C-section beam. Suspension was totally new with front and rear fiberglass transverse monoleaf springs. Aluminum and other lightweight materials were used wherever possible to cut unwanted pounds.

Sixteen-inch cast-aluminum wheels measured a half-inch wider in back. The four-wheel disc brakes had semi-metallic linings and aluminum calipers. The engine was a 205 horsepower L83 Cross Fire 5.7-liter V-8. A choice was offered between a four-speed automatic or 4+3 Doug Nash manual (with overdrives in the top three gears). All this and more was standard with the 1984 Corvette. More than 51,000 were sold during the extended production run and kicked off the third generation in grand fashion.

Two years later, a convertible returned to the Corvette lineup after a nine-year hiatus, and it appeared just in time to pace the 70th running of the Indy 500 on May 25, 1986. For the second time, a collection of high-profile Corvette Indy Pace Car replicas was marketed to the public. A third Corvette paced the lead lap at Indianapolis in 1995 and spawned yet another group of Pace Car replicas.

Late in 1986, the first in a series of tough Corvettes appeared on the lot of a New Jersey Chevrolet dealership founded by Malcolm Konner. Konner Chevrolet put together 50 of its "Malcolm Konner Commemorative Edition" Corvettes, and one was retrofitted with a twin-turbocharged engine supplied by Callaway Cars, Inc. Reeves Callaway, Callaway Cars' founder, had toyed with aftermarket turbo modifications of various models, from BMW to Volkswagen, since 1977, but the Konner installation was just a stepping stone towards Callaway's biggest break—an agreement with Chevrolet to build the Callaway Twin Turbo Corvette.

Chevrolet officially assigned the Callaway package an RPO code, B2K, in June 1986. Production began the following month and the Twin Turbo

but various stumbling blocks delayed that debut, which would have been in the fall of 1982. While 43 pre-production, third-generation 1983 Corvettes were built, none were released to the public. Chevrolet skipped over the 1983 model and introduced its next generation Corvette as a 1984 with an extended production run.

Everything about the car was new, from its modern chassis to its roomier interior and restyled body. The 1984 Corvette shell, created by Jerry Palmer, GM designer, was state-of-the-art in both form and function. Its drag coefficient was 0.34, down nearly 25 percent in comparison to the body that was left behind in 1982. Beneath that shell was an innovative birdcage structure integrated with a backbone-type frame that mounted the drivetrain

Four cams, 32 valves, and an all-aluminum construction qualified the ZR-1's LT5 V-8 as the most exotic production Corvette powerplant ever. Rated at 375 horsepower from 1990 to 1992, the LT5 was pumped up to 405 horses for 1993-95.

killer Corvette debuted as a 1987 model. Chevrolet shipped fully assembled 1987 Corvettes to Callaway Cars where they were converted into Twin Turbos. Although the price for the option alone was $19,995, those in the need for speed couldn't have cared less. With 345 horsepower, a 1987 Callaway Corvette could reportedly hit 177 miles per hour.

There were 184 built that first year. RPO B2K stayed on the Corvette options list through 1991.

Back in the regular-production world, Chevrolet marked a fiberglass birthday in 1988 with a special 35th anniversary Corvette. Like the first Corvette in 1953, the 35th anniversary model came only in white. White leather sport buckets with anniversary headrest embroidery and a commemorative console plaque were also part of the deal. Power seats, air conditioning, a sport handling package, and external identification were also included. The price for the package was $4,975, and 2,050 were built.

The legendary "King of the Hill," the ZR-1, debuted as a 1990 model after the public was teased with a prototype introduction in 1989. The heart of this beast was the 375 horsepower LT5 5.7-liter V-8, engineered by Lotus in England and built by Mercury Marine in Stillwater, Oklahoma. Dual overhead cams, four valves per cylinder, and an all-aluminum construction were just the beginning of the LT5's innovative appeal.

As for the rest of the package, ZR-1 brakes were huge—13 inches in front, and 12 in the rear. In order to house the equally huge Goodyear Eagle 315/35ZR-17 GS-C rear tires required to handle all that power, the ZR-1 received an exclusive widened tail section with unique square taillights. ABS and Z51 suspension were also included.

Whether in the curves or flat-out, the ZR-1 was a world-class street killer. Its time-honored 0 to 60 clocking was 4.9 seconds. Quarter-mile performance came in at 13.4 seconds with a 108-mile-per-hour trap speed. The top end was nearly 180 miles per hour. All this dominating power did come at a price. ZR-1 equipment added around $25,000 to the $32,000 normal asking price for a 1990 Corvette coupe. While that figure didn't deter diehards at first, it did limit the car's appeal. Actual production never did reach projections, not even after Chevrolet leapfrogged the 1993 ZR-1 over the Viper as the most powerful car in

Yet another birthday present to Corvette customers arrived in 1993, and an exclusive paint scheme was once more part of the package.

Callaway Cars' outrageous Twin Turbo Speedster of 1991 was designed by Paul Deutschmann. A 450 horespower, 5.7 liter small-block was beneath those twin hood scoops. The original price for this beast was $150,000.

The rarest of the 1993 40th anniversary Corvettes was the ZR-1 coupe. Only 245 were built. This particular 40th anniversary ZR-1 was modified by aftermarket builder Doug Rippie Motorsports. Only eight ZR-1s received DRM's emissions-legal Black Widow touch.

Chevrolet introduced this eye-catching Copper Metallic finish for the Corvette in 1994, only to find it quite difficult to apply evenly. Because of this, the shade was discontinued early in the run. Reportedly, only 110 1994 Copper Metallic Corvettes were released.

America by upping the LT5 ante to a whopping 405 horsepower.

After 3,049 ZR-1 Corvettes were built the first year, production rapidly declined as the gleam wore off; even more so after Chevrolet introduced its second-generation LT1 in 1992. ZR-1 production dropped to 2,044 for 1991, 502 for 1992, and 448 for 1993, 1994, and 1995. According to Chevrolet General Manager Jim Perkins, both the success of the much less expensive LT1 Corvette and the rising costs associated with continuing the ZR-1's limited production run helped spell the end for the King of the Hill.

After taking away a bit of the ZR-1's exclusive outward appeal by adding the widened tail to all Corvettes in 1991, Chevrolet really knocked the knees out from under the King of the Hill in 1992. A truly hot pushrod small-block, the second-generation LT1, was introduced that year. It produced 300 horsepower and was the latest in Chevy's long line of 5.7-liter small-blocks. The LT1 went a long way and tempted Corvette buyers to save their $25-30,000 and stick with standard Corvette performance. A 1992 LT1 could hit 60 miles per hour in 5.7 seconds, while the quarter-mile went by in 8.4 ticks. Top end was listed at a tad more than 160 miles per hour. LT1 muscle was just what the doctor ordered to revive the Corvette's spirits going into its fourth decade on the road.

To mark that special anniversary, Chevrolet once again rolled out a birthday edition. Wearing a price tag of $1,455, RPO Z25—Chevrolet's 40th anniversary package—was available on all 1993 Corvettes, LT1 coupes and convertibles, and the brawny ZR-1. Exclusive Ruby Red metallic paint with matching leather inside, color-keyed wheel centers, 40th anniversary logos, and chromed emblems for the hood and fuel-filler door were also

On August 28, 1995, the last of Chevrolet's 6,938 ZR-1 Corvettes rolled off the Bowling Green assembly line before a crowd of company officials and press and signaled the end of the six-year run for the King of the Hill.

Chevrolet offered special Collector's Edition models in 1996 (foreground) and 1982. Both marked the end of a design generation.

The new LT4 small-block is standard for the 1996 Grand Sport and optional on all other Corvettes. Its 30 additional horses, compared to the standard 300 horsepower LT1, will be warmly welcomed.

thrown in as part of the deal. Additional special identification of the "40th" logo was embroidered on the Ruby Red leather bucket seats' headrests. There were 4,333 coupes, 2,171 convertibles, and 245 ZR-1's of this special edition produced.

Even after 40 years, America's sports car remained strong. Improvements to the LT1's induction gear helped the 1994 Corvette feel even stronger than the 1993, even though advertised output remained at 300 horsepower. Minor overall improvements followed, and the LT4 was introduced in 1996. Like its LT1 cousin, the 5.7-liter LT4 V-8 had aluminum heads, roller lifters, and sequential-port fuel injection. Various improvements included a more aggressive cam; better breathing

heads; upgraded, high-flowing injection; and 10.8:1 compression (up from the LT1's 10.4:1). The LT4's output was 330 horsepower.

The LT4 was available at extra cost in all 1996 Corvettes, but was the standard powerplant for the Grand Sport special edition models. The newest Grand Sport was inspired by Duntov's five lightweight Grand Sport racers of 1963 and featured an exclusive competition-type paint scheme. All of the 1,000 Grand Sport coupes and convertibles built for 1996 were painted Admiral Blue Metallic with white racing stripes and red "Sebring-style" hash marks on the driver's side front fender. Various bits of special identification, 17-inch black aluminum wheels, black brake calipers with bright Corvette lettering,

Painted to revive memories of its racing namesakes from earlier days, the 1996 Grand Sport is offered in coupe and convertible form.

and special bucket seats were part of the package, which was listed under RPO Z16. Grand Sport coupes received P275/40ZR-17 rubber up front and P315/35ZR-17 in the rear, with a pair of fender flares added in back to help house the wider tires. Grand Sport convertible tires were P255/45ZR-17 (front) and P285/40ZR-17 (back).

A second special edition Corvette for 1996 also followed in another earlier model's tracks. Like its 1982 counterpart, the 1996 Collector's Edition arrived just in time to salute the last of the latest generation of Corvettes before the next all-new model arrived. Exclusive Sebring Silver paint, special identification, silver 17-inch aluminum wheels, black brake calipers with bright Corvette lettering, and "Collector Edition" embroidery inside helped create what Chevrolet officials called an "eminently collectible" offering. As with base models, the 1996 Collector's Edition came standard with the LT1 V-8, and the LT4 was an option. Collectible or not, the 1996 Collector's Edition served as a noteworthy send-off for the C4 Corvette.

6 1997–Present
The Best 'Vettes Yet

Originally conceived in 1988 with hopes for an August 1992 introduction, the long-awaited C5 Corvette finally debuted before the American public in January 1997. Various stumbling blocks, which involved a massive wave of red ink at General Motors during the 1990s, kept pushing back the new model's release date. For a while, the project even teetered on the brink of cancellation. In the end, the wait was more than worth it.

Both critics and company officials alike couldn't say enough about the C5. "The fifth-generation Corvette is a refined Corvette in all the right ways," bragged chief engineer David Hill in 1997. "It's more user-friendly, it's easier to get in and out of, and it's more ergonomic. It has greater visibility; it's more comfortable and more functional. It provides more sports car for the money than anything in its market segment. It'll pull nearly 1g, and it starts and stops quicker than you can blink."

There were many changes to the fifth-generation Corvette. Underneath that slippery wind-tunnel-tested skin was a totally redesigned frame based on a rigid center tunnel spine flanked by hydroformed perimeter rails. The innovative hydroformed side rails were created using immense water pressure to turn steel tubes into super-strong rectangular sections, with the result being a foundation that was 4.5 times stronger than its C4 predecessor. This extra rigidity helped improve ride and handling and did away with many of the squeaky gremlins inherent in earlier Corvettes.

The C5's rear-mounted transmission also assisted ride and handling. It was a long-discussed idea that brought weight distribution closer to the preferred 50/50 balance. This new layout also freed up space beneath the passenger compartment, which meant the driver and passenger had more room to stretch out and ride comfortably. Entry and exit was enhanced thanks to the strengthened frame rails that traded excess mass for a lower sill height, which was down 3.7 inches.

Bigger and better brakes and tires and a more sophisticated suspension were part of the deal, as was the new LS1 Gen III V-8, Chevrolet's first all-aluminum regular-production small-block. Although it shared its 5.7-liter displacement, two-valves-per-cylinder layout, and 4.4-inch bore centers with previous pushrod small-blocks, the LS1 represented a marked departure from its LT1 forerunner. "The new LS1 has the simplicity and compactness of the pushrod layout, but with porting so efficient and a valvetrain so light and stiff, it breathes like an overhead-cam motor," bragged Hill. Output was 345 horsepower.

The C5 Corvette was designed first as a convertible, but debuted in 1997 only in coupe form. Its platform was the strongest yet.

The C5 Corvette's LS1 V-8 became Chevrolet's first all-aluminum small-block in 1997. The output was 345 horsepower.

Initially, the C5 was only offered as a targa top coupe, but a convertible joined the ranks in 1998 and quickly copped Motor Trend's coveted Car of the Year trophy. The new droptop was chosen as the fourth Corvette to pace the 1998 Indianapolis 500, an honor that happened again for another C5 in 2003.

The C5 convertible's trunk was new for 1998, and the Corvette's first since the Sting Ray superseded the solid-axle cars in 1963. A clever dual-compartment gas tank, deletion of a spare tire (by making extended-mobility run-flat tires standard) and the compact, non-power aspects of the C5 convertible's manual folding top all worked in concert to make a trunk possible. The C5's warmly welcomed rear cargo area measured 11.2 cubic feet in top-down mode; 13.9 with the roof unfolded and in place. Two bags of golf clubs would fit in there, which was the example Chevrolet's promotional people used to push the point of how spacious the storage compartment was.

A trunk also appeared on the new fixed-roof hardtop C5, which debuted in 1999 to expand the Corvette lineup to three different models. Initially planned as a low-buck alternative, the Corvette hardtop showed up with a bit more standard equipment than originally envisioned, yet it remained the cheapest of the line with a base price of around $38,800, nearly $500 less than the targa top sport coupe.

A C5 convertible debuted for 1998, along with the Corvette's first trunk since 1962.

Chevrolet dusted off the Z06 label for another hotter-than-hot Corvette model in 2001. Special wheels and braking cooling ducts were part of the latest, greatest Z06 package.

Powering the 2001 Z06 was the LS6 small-block, which was rated at 385 horsepower. Displacement was the same as the LS1: 5.7 liters, or about 350 cubic inches.

That entry-level hardtop model became the base for yet another new Corvette, which was (in Chevrolet's words) "aimed directly at diehard performance enthusiasts at the upper end of the high-performance market." Introduced for 2001, the Z06 borrowed its name from the batch of race-ready Sting Rays built in 1963—an appropriate choice considering the two cars shared the same ideal. As Corvette brand manager Jim Campbell explained, "The new Z06 will have great appeal for those who lust after something more—that indefinable thrill that comes from being able to drive competitively at 10/10ths in a car purpose-built do to exactly that."

Zora Arkus-Duntov's original Z06 package included the Sting Ray's hottest V-8 that worked in concert with a beefed-up suspension and brakes to help make a trip from the showroom right to the racetrack possible. The plan was much the same in 2001. Wider wheels and tires, special rear and front brake-cooling ductwork, and the exclusive FE4 suspension, which featured a larger front stabilizer bar, a stiffer leaf spring in back, and revised camber settings at both ends were standard for the reborn Z06. Weight throughout the Z06 was cut by about 100 pounds overall compared to a 2001 Corvette sport coupe.

The Z06's exclusive wheels were 17x9.5 inchers in front, 18x10.5 in back—in both cases, one inch wider than the standard C5 rims. Goodyear Eagle F1 SC tires, P265/40ZR-17 in front and P295/35ZR-18 in back, were mounted on the Z06's widened rollers. The C5s in 2001 featured Eagle F1 GS rubber, P245/45ZR-17 at the nose and P275/40ZR-18 at the tail.

Beneath the new Z06's hood was another hot power source, this one created only for this application, which was named using a legendary options code, LS6, from Corvette days gone by. In 1971, the LS6 454 big-block was the hottest Corvette V-8 offered that year. It was the same deal for the 2001 LS6, which was based on the C5's existing LS1 small-block. A recast block, stronger pistons, raised compression (from 10.1 to 10.5:1), a lumpier cam, and bigger injectors were just a few of the dozens of LS6 improvements. Output was 385 horsepower, which was 40 more than the LS1. A new M12 six-speed manual transmission, the only gearbox available for the Z06, was behind the LS6.

These new parts and many others helped make the Z06, in Chevrolet's words, "simply the quickest, best handling production Corvette ever." David Hill added, "We've enhanced Corvette's performance persona and broken new ground with the new Z06. With 0 to 60 [times] of four seconds flat, and more than 1g of cornering acceleration, the Z06 truly takes Corvette performance to the next level. In fact, the Corvette Team has begun referring to it as the C5.5, so marked are the improvements we've made and the optimization of the car in every dimension."

Chevrolet sold 5,773 Z06 Corvettes in 2001, followed by another 8,297 in 2002, the year LS6 output soared to 405 horsepower. Continued popularity in 2003 proved the Z06 was no imposter.

As for Corvette popularity in general, it remained strong in 2003. To mark 50 years of customer loyalty, Chevrolet again created a special anniversary model. All 2003 Corvettes wore 50th Anniversary badges, but for an extra $5,000, a coupe or convertible could be dressed up further with the 1SC package, a cosmetic addition that wasn't offered to Z06 buyers.

Included in the 1SC deal were unique fender emblems, Anniversary Red Xirallic crystal paint, champagne-colored aluminum wheels, and a Shale interior. Embroidered 50th Anniversary logos appeared on the floor mats and headrests, and convertibles were additionally treated to Shale-colored soft tops. All 50th Anniversary Corvettes were equipped with Magnetic Selective ride control, which was yet another high-tech approach to keep the car's dirty side down.

Commemorating a half century worth of Corvette history was a heavily anticipated happening, but the moment was somewhat over-shadowed by news of what awaited America's sport car in the near future. Those 50 candles had barely stopped smoking when all attention turned to the upcoming C6, slated for a 2005 debut. By late 2003, the revamped sixth-generation Corvette was dominated magazine covers in preparation for its official public unveiling at the Detroit Auto Show in January 2004. Again, the plan is to roll out a sport coupe C6 first later in the year, followed by a convertible running mate in the fall of 2005. The fiberglass faithful was already buzzing well before the debut in Detroit.

"The C6 represents a comprehensive upgrade to the Corvette," explained David Hill late in 2003. "Our goal is to create a Corvette that does more things well than any performance car. We've thoroughly improved performance and developed new features and capabilities in many areas, while at the same time, systematically searching out and destroying every imperfection we could find."

Apparently, hideaway headlights, a Corvette trademark since 1963, qualified as imperfections. They were deleted from the new body in favor of exposed lamps that give the car, in many opinions, a distinct Viper-esque appearance. The new look, though, is certainly hot and further enhanced by softly rounded contours that lessen the sharp impact made by the C5. Technically speaking, with a drag coefficient of 0.28, the C6 body ranks as the most aerodynamic Corvette shell to date. It also measures five inches shorter and about an inch narrower than its C5 forerunner.

Beneath that super slick skin is a new power source, the 6.0-liter LS2 Gen IV V-8, the largest, most powerful small-block ever unleashed by Chevrolet. At the same time, the 400 horsepower LS2 also ranks as the most fuel-efficient performance engine found in the world's top sports cars. The combined city/highway fuel economy estimate for the 2005 Corvette is 22.6 mile per gallon, a major plus considering the direction gas prices have been taking during recent months.

Eighteen-inch wheels bring up the C6's nose, while 19-inchers follow in back. The same basic C5 chassis carries over, but with significant enhancements to further improve ride comfort and handling precision. The brakes are enlarged along with the wheels, and most suspension components have been replaced by superior pieces.

All told, the C6 model easily qualifies as the latest "best 'Vette yet." How many more better generations remain in the Corvette's future is anyone's guess. Here's to another 50 years.

Chevrolet's C5-R Corvette competition team dominated IMSA racing's American Le Mans Series (ALMS) for the second straight year in 2002, and captured 9 GTS class wins in 10 races. Here, the No. 3 C5-R, driven by Ron Fellows, Johnny O'Connell, and Frank Freon, heats things up at Road Atlanta during the final ALMS event in 2002.

Index

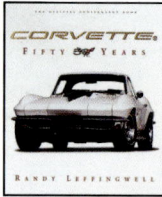

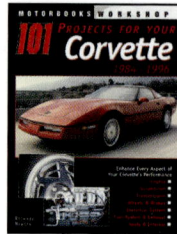

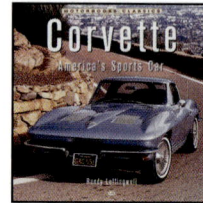

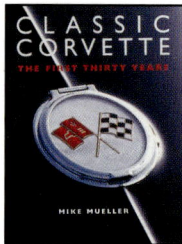

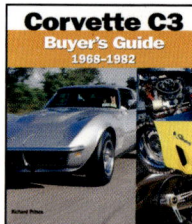

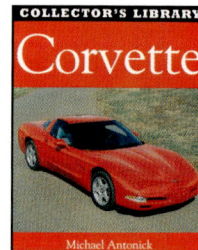

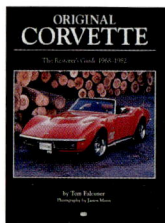

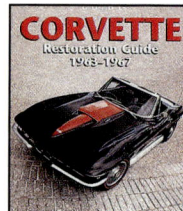

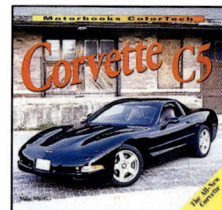